AF270474

EXPLORE BIOMES

RAIN FOREST BIOMES

BY CECILIA PINTO McCARTHY

Kids Core

An Imprint of Abdo Publishing
abdobooks.com

Published by Abdo Publishing, a division of ABDO, PO Box 398166, Minneapolis, Minnesota 55439. Copyright © 2024 by Abdo Consulting Group, Inc. International copyrights reserved in all countries. No part of this book may be reproduced in any form without written permission from the publisher. Kids Core™ is a trademark and logo of Abdo Publishing.

Printed in the United States of America, North Mankato, Minnesota.
052023
092023

Cover Photo: Kevin Wells/iStockphoto
Interior Photos: iStockphoto, 4–5; Joe McDonald/The Image Bank/Getty Images, 6; Shutterstock Images, 8, 13, 21, 24, 28 (toucan, bat, orangutan), 29 (jaguar); OG Photo/iStockphoto, 10–11; Blue Ring Media/Shutterstock Images, 12, 28–29; Manta Photo/iStockphoto, 14; Ron Crabtree/Photographer's Choice RF/Getty Images, 16; Enrico Pescantini/Shutterstock Images, 18–19; MerlinTuttle.org/Science Source, 22; Vaclav Sebek/Shutterstock Images, 25; Juan Carlos Vindas/Moment/Getty Images, 26; Margaret Jone Wollman/Shutterstock Images, 28 (frog)

Editor: Angela Lim
Series Designer: Ryan Gale

Library of Congress Control Number: 2022949094

Publisher's Cataloging-in-Publication Data

Names: McCarthy, Cecilia Pinto, author.
Title: Rain Forest biomes / by Cecilia Pinto McCarthy
Description: Minneapolis, Minnesota: Abdo Publishing Company, 2024 | Series: Explore biomes | Includes online resources and index.
Identifiers: ISBN 9781098291129 (lib. bdg.) | ISBN 9781098277307 (ebook)
Subjects: LCSH: Rain forests--Juvenile literature. | Biotic communities--Juvenile literature. | Habitats--Juvenile literature. | Life zones--Juvenile literature. | Rain forest animals--Juvenile literature. | Rain forest plants--Juvenile literature. | Rain forest ecology--Juvenile literature.
Classification: DDC 577--dc23

CONTENTS

Agoutis eat mainly fallen fruit and nuts.

CHAPTER 1

SPREADING SEEDS

An agouti searches the rain forest floor in South America. It is looking for Brazil nut fruit pods. The pods have a thick shell that most animals cannot break open. Agoutis are one of only a few animals that can open the pods.

Brown capuchins use stones as tools to crack the hard shells of nuts.

The agouti finds a pod and gnaws a hole in the shell. The pod has seeds inside. The agouti eats a few seeds. It carries other seeds away and buries them in different places. The next day, the agouti returns. It digs up seeds to eat. The agouti forgets where some seeds are buried. The uneaten seeds will grow into new Brazil nut trees.

What Is a Rain Forest Biome?

Earth is made up of biomes. Each biome has certain kinds of animals, plants, and nonliving things. Agoutis are rain forest animals. Brazil nut trees are one of the many plants species found in rain forests. Nonliving things in a biome include rocks and water.

Rain Forest Products

Rain forests are sources of wood, food, and medicines for people. Avocados, cocoa, and bananas are some of the foods that come from rain forest plants. Many plants have chemicals that are used in medicine. For example, medicine made from the rosy periwinkle helps treat cancer. Quinine is a drug made from the bark of the cinchona tree. It is used to treat malaria.

Climate is another part of a biome. Climate describes the weather in an area over a long period of time. Rain forests receive heavy rainfall. Some get up to 394 inches (1,000 cm) each year.

Rain forest biomes help the environment.
Plants take in a gas called carbon dioxide.
High levels of carbon dioxide add to climate
change. This is a long-term shift in weather
patterns, such as rising temperatures. Because
there is much plant life in rain forests, these
biomes play a major role in reducing the effects
of climate change.

Explore Online

Visit the website below. Does it give
any new information about rain forest
biomes that wasn't in Chapter One?

Rain Forest

abdocorelibrary.com/rain-forest
-biomes

Scarlet macaws can be found in large groups in the canopy layer of a rain forest.

TYPES OF RAIN FOREST BIOMES

Rain forests can be divided into layers. The tallest trees form the emergent layer. Emergent trees have few leaves near the base of their trunks. Most of their **foliage** is at the top, where leaves can catch the most sunlight.

Rain Forest Layers

Plants of different heights make up the layers of a rain forest biome. They provide shelter for many animals.

The canopy is beneath the emergent layer. It blocks sunlight, rain, and wind from the lower areas of the forest.

The understory is below the canopy. This area is darker and more **humid** than

the canopy. Shrubs grow in this layer. The bottom layer is the forest floor. It is dark and covered with dead plant material.

Tropical and Temperate Rain Forests

Rain forests are found on every continent except Antarctica. There are two types of rain forests. They are tropical rain forests and temperate rain forests. They are found in different parts of the world.

The Amazon River runs through the Amazon rain forest.

Tropical rain forests are located near the **equator**. They receive steady amounts of sunlight. Tropical rain forests are warm and rainy year-round. They are hot and humid.

Average temperatures range from 70 to 86 degrees Fahrenheit (21–30°C).

The Amazon rain forest is the largest tropical rain forest. It contains about 40,000 species of plants and approximately 16,000 types of trees. It is also home to thousands of species of birds and fish and about 2.5 million insect species.

Transpiration

Plants give off water vapor through their leaves. This process is called transpiration. The water vapor forms clouds. The clouds then release rain. A tree in the Amazon rain forest can give off about 264 gallons (1,000 L) of water vapor every day through transpiration. Air currents move this moisture around the planet.

Olympic National Park in Washington protects the Hoh Rain Forest. Some trees in this temperate rain forest are more than 1,000 years old.

Temperate rain forests are farther from the equator. They are cooler than tropical rain forests. Winters tend to be mild. Temperatures stay between 50 and 70 degrees

Fahrenheit (10–21°C) throughout the year. Many temperate rain forests are found in coastal regions. Moist air blows inland from the ocean.

Temperate rain forests run along the Pacific coast of North America. These forests begin in California and stretch north through Canada and into Alaska. The Tongass National Forest in Alaska is a temperate rain forest. It is the largest national forest in the United States.

Further Evidence

Look at the website below. What evidence does it give to support Chapter Two?

Rain Forest Habitat

abdocorelibrary.com/rain-forest
-biomes

Sloths move very slowly. They travel only about 41 yards (37 m) a day.

CHAPTER 3

LIFE IN RAIN FOREST BIOMES

The layers of a tropical rain forest are home to many living things. Birds such as the harpy eagle and hornbill live in the emergent layer. Most animals live in the canopy. Orangutans, toucans, and sloths live in this layer.

The canopy is also rich with plants that provide food and shelter. Shrubs and plants that do not need much sunlight grow in the understory. Jaguars hunt in this layer. On the forest floor, slugs and fungi break down dead matter. They return **nutrients** to the soil.

Temperate rain forests have different plants and animals than tropical rain forests. Trees such as redwoods, spruces, and hemlocks are common in temperate rain forests. Northern spotted owls nest in the trees. They hunt flying squirrels and other small animals. Salmonberry shrubs grow in the understory. Mosses and ferns blanket the ground. Banana slugs, salamanders, bobcats, and elk also make their homes in temperate rain forests.

Living Together

Plants and animals need one another to survive. Animals eat plants. They also use plants for shelter and to raise their young. Bromeliad plants grow on tree trunks in tropical rain forests. Rainwater collects in the center of these plants and forms a pool. Female strawberry poison dart frogs carry their young to these pools. The young are called tadpoles. Tadpoles stay in the pools until they become adults.

Animals **pollinate** plants. The tube-lipped **nectar** bat lives in the tropical rain forests of South America. It uses its tongue to reach nectar at the bottom of flowers. As it drinks, pollen from the flower rubs onto the bat's head. The bat travels to another flower. It drops pollen into the flower as it feeds. This helps the plants make seeds.

Some animals and birds eat fruits and nuts. The seeds are dropped on the ground in animal waste. Red crossbill birds eat pine cone seeds in the rain forests of the Pacific coast. They help spread seeds to new areas.

A Shrinking Biome

Rain forest biomes are shrinking. People cut down trees for wood and to grow crops. Fewer trees mean less shelter and food for animals. Climate change threatens rain forest plants and animals. People are working to manage and save these biomes.

Kermode bears are also called spirit bears. Despite their white color, they are a type of black bear.

Animals also need each other. Some animals are **predators**. The Great Bear Rain Forest in Canada is a temperate rain forest. There, Kermode bears fish for salmon. They carry the fish into the forest to eat. Leftover bits of fish rot and add nutrients to the soil. This helps trees and other plants grow. Snails and slugs also feed on the dead fish.

Leaf-cutter ants can carry leaves that are 50 times
their weight. They use leaves to help grow fungi
to eat.

Rain forests are home to many animals. The high amount of rainfall allows thousands of plants to grow. These biomes are important to people too. They provide food and oxygen for the world.

Dr. Oliver Moore is a scientist with the Alliance for Scotland's Rainforest. He explains the importance of Scotland's rain forests.

> [Temperate rain forests] are . . . fascinating and full of rare species of plants, with new species being discovered all the time.

Source: Vivien Cumming. "Britain Has a Rain Forest." *National Geographic*, 21 Apr. 2022, nationalgeographic.co.uk. Accessed 21 Oct. 2022.

What's the Big Idea?

Read this quote carefully. What is its main idea? Explain how the main idea is supported by details.

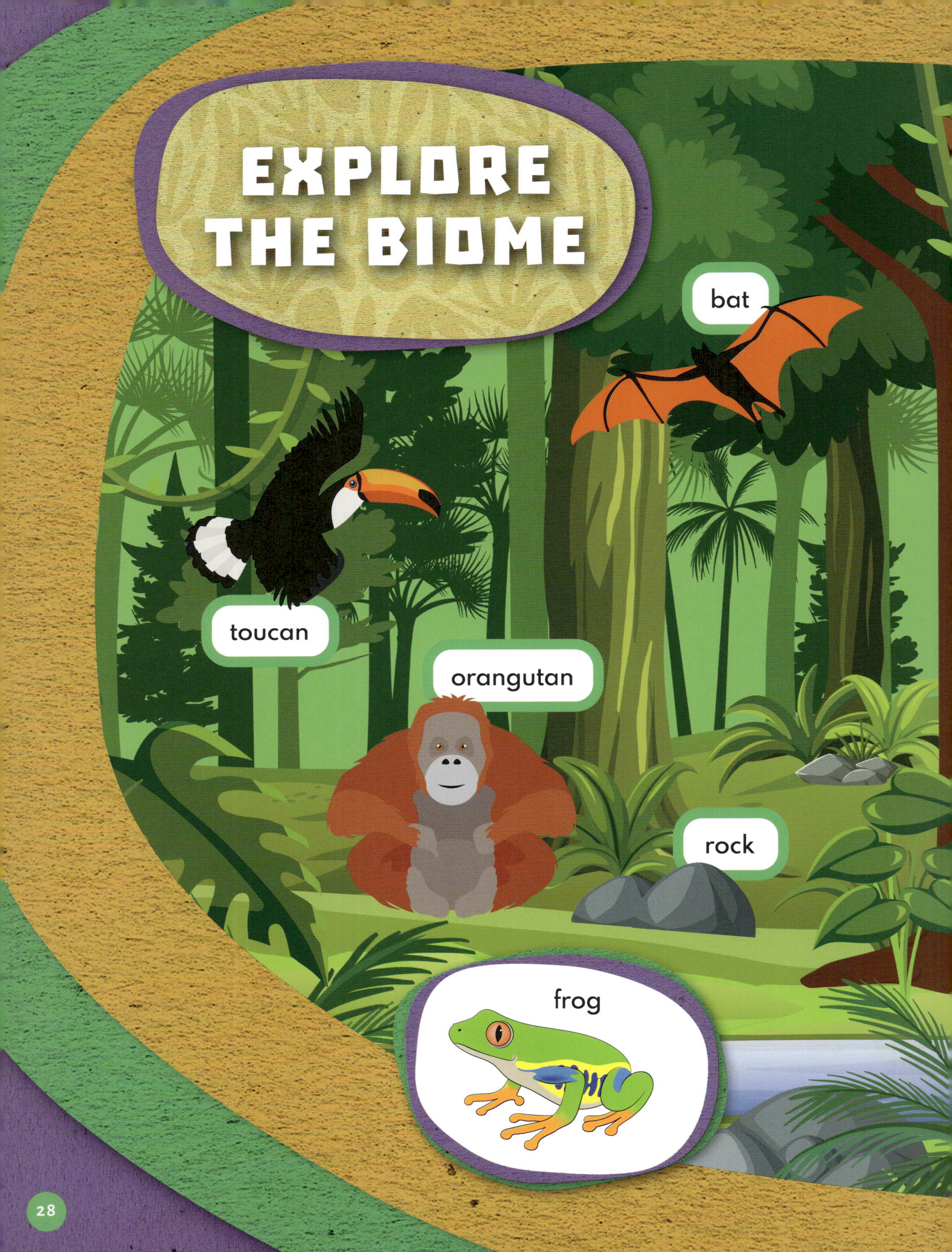
EXPLORE
THE BIOME
bat
toucan
orangutan
rock
frog
28

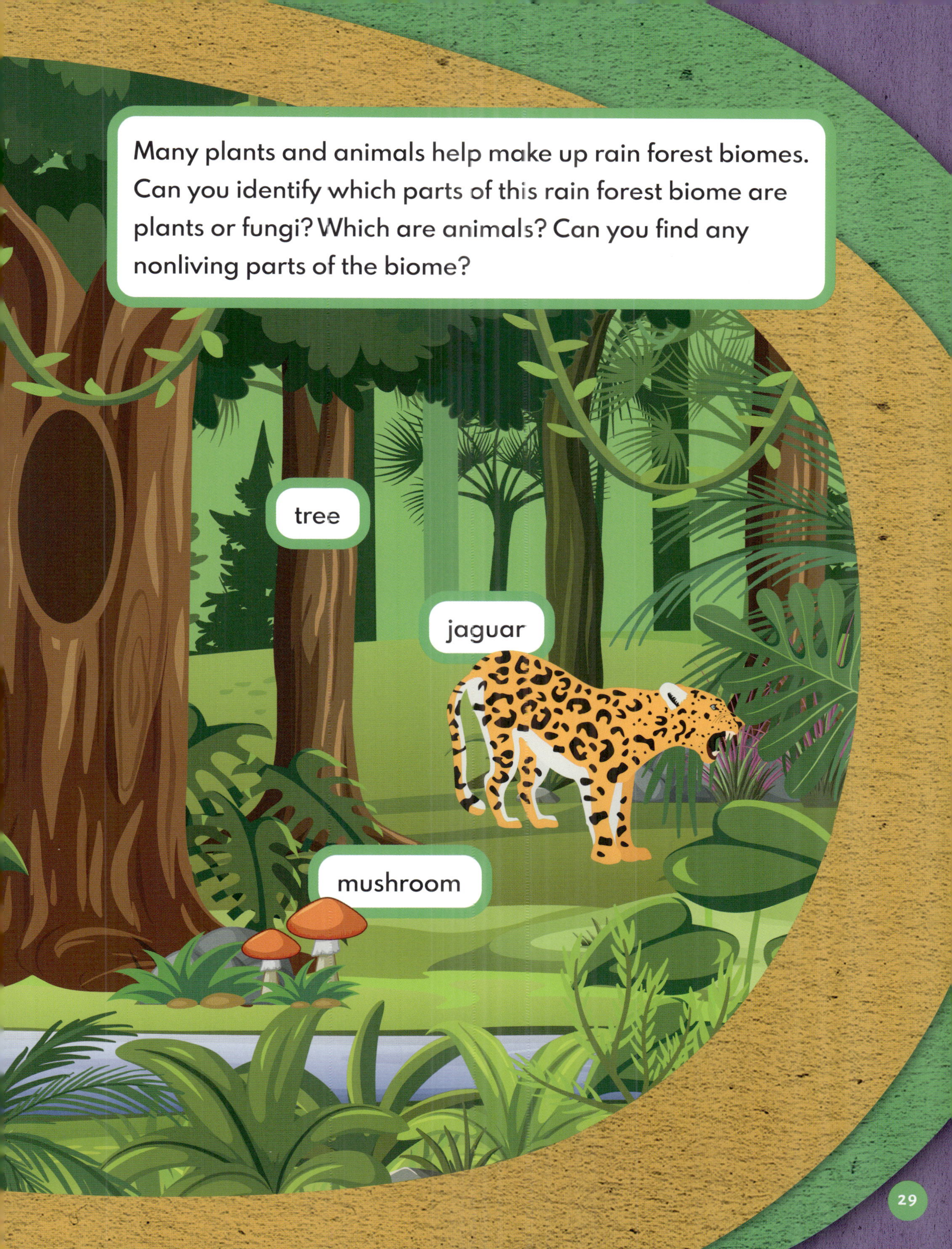

Many plants and animals help make up rain forest biomes. Can you identify which parts of this rain forest biome are plants or fungi? Which are animals? Can you find any nonliving parts of the biome?
tree
jaguar
mushroom
29

Glossary

equator
the imaginary circle that goes around Earth halfway between the North and South Poles

foliage
leaves on a tree or other plant

humid
damp or moist

nectar
a sweet liquid that plants make

nutrients
materials that living things need in order to grow and stay alive

pollinate
to move pollen to a plant to fertilize it

predators
animals that hunt other animals for food

Online Resources

To learn more about rain forest biomes, visit our free resource websites below.

Visit **abdocorelibrary.com** or scan this QR code for free Common Core resources for teachers and students, including vetted activities, multimedia, and booklinks, for deeper subject comprehension.

Visit **abdobooklinks.com** or scan this QR code for free additional online weblinks for further learning. These links are routinely monitored and updated to provide the most current information available.

Learn More

London, Martha. *Ecosystems.* Abdo, 2022.

Munro, Roxie. *Anteaters, Bats & Boas.* Holiday House, 2021.

Ridley, Sarah. *Who Ate the Butterfly?* Crabtree, 2020.

Index

About the Author

Cecilia Pinto McCarthy enjoys sharing her love of nature with young readers. When she is not writing, she teaches at a nature center. She lives north of Boston, Massachusetts.